GALAXIES

Rebecca Woodbury, Ph.D., M.Ed.

Gravitas Publications Inc.

GALAXIES

Illustrations: Janet Moneymaker

Galaxies
ISBN 978-1-950415-46-5

Published by Gravitas Publications Inc.
Imprint: Real Science-4-Kids
www.gravitaspublications.com
www.realscience4kids.com

RS4K

Photo credits: Cover & Title Pg: NASA, ESA, and the Hubble Heritage Team (STScI/AURA) Acknowledgement: William Blair (Johns Hopkins University); Above, NASA, ESA, and The Hubble Heritage Team/STScI/AURA; P.3. NASA's James Webb Space Telescope, CC BY SA 2.0; P.5. NASA/JPL-Caltech/R. Hurt (SSC_Caltech); P.9. ESA/Hubble & NASA, J. Dalcanton, Dark Energy Survey/DOE/FNAL/NOIRLab/NSF/AURA; Acknowledgement: L. Shatz; P.13: Top, NASA/JPL-Caltech/R. Hurt (SSC/Caltech); Bottom, Public Domain; P.15. 1. Pablo Carlos Budassi, CC BY SA 4.0; 2. NASA, ESA, S Beckwith (STScI), and The Hubble Heritage Team-STScI/AURA; 3. NASA, ESA, and The Hubble Heritage Team (STScI/AURA); P.17. ESA/Hubble & NASA, Acknowledgement: JudySchmidt-JBlakeslee, Dominion Astrophysical Observatory, Science Acknowledgement: MCarollo/ETH-Switzerland; P.19. NASA, ESA, and The Hubble Heritage Team/STScI/AURA; P.20-21. NASA, ESA, Adam G. Riess/STScI

A **galaxy** is a huge group
of **stars, planets,** gas, dust,
comets, asteroids, and
other objects in space.

Wow! There's a lot
of stuff in a galaxy!

A spiral galaxy

Our Sun and Earth are part
of the **Milky Way Galaxy.**

Oooh! We live in a candy bar!

What a silly mouse!

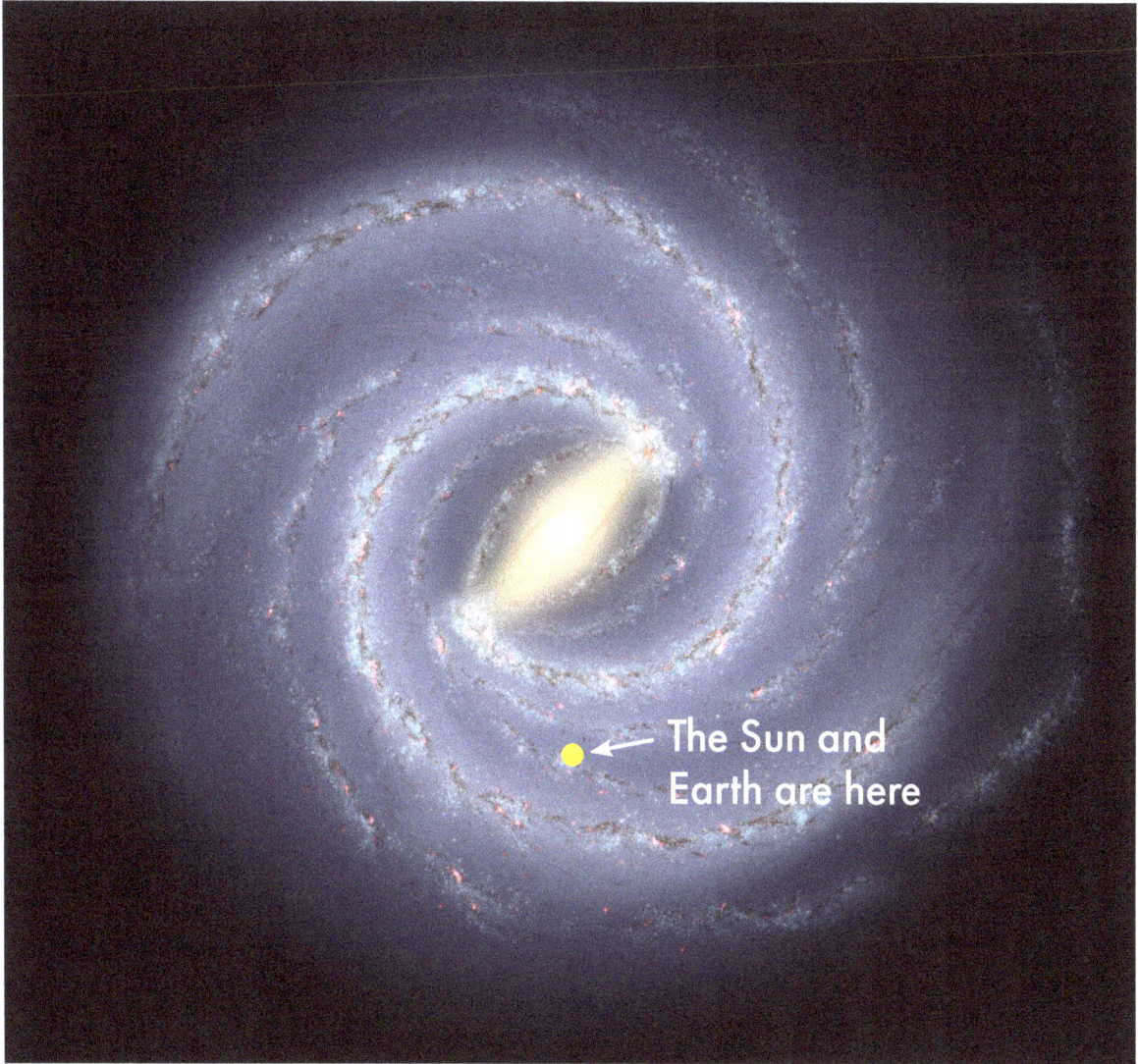

The Sun and Earth are here

The Milky Way Galaxy

Astronomers cannot tell exactly what the Milky Way Galaxy looks like. It is so huge that we have no way to see it from the outside. Rockets cannot fly high enough to get outside the Milky Way Galaxy.

Bye!

Have a good trip!

Studying other galaxies is one way astronomers use to guess what the Milky Way Galaxy looks like.

Astronomers also study how the stars are arranged in our galaxy.

So many galaxies to study!

A group of different types of galaxies

Astronomers use **telescopes** to view other galaxies. A telescope is a tool that makes objects in space look bigger.

Galaxies!

We think the Milky Way Galaxy is shaped like a pinwheel. It has a bulge made of tightly packed stars in the center and arms made of stars that reach out in a spiral.

Look at my mini galaxy!

It's just a pinwheel!

Yes. But it shows the idea of a galaxy.

Milky Way Galaxy – top view

Milky Way Galaxy – side view

The Milky Way is called a **spiral galaxy** because it has spiraling arms.

Astronomers can see many other spiral galaxies in the universe.

I wonder if anyone lives out there.

Maybe.

Different types of spiral galaxies

Another type of galaxy is the **elliptical galaxy.** An elliptical galaxy is made of tightly packed stars and has no arms. It can look like a huge star.

It looks like a giant fuzzball.

I have those under my bed.

An elliptical galaxy

There are also **irregular galaxies.**
An irregular galaxy can be any shape.

Beautiful!

An irregular galaxy

Space is filled with billions and trillions of beautiful galaxies!

How to say science words

asteroid   (AA-stuh-royd)

astronomer   (uh-STRAH-nuh-muhr)

comet   (KAH-muht)

Earth   (ERTH)

elliptical   (ih-LIP-tih-kuhl)

galaxy   (GAA-luhk-see)

irregular   (ih-REH-guh-luhr)

Milky Way Galaxy   (MIL-kee WAY GAA-luhk-see)

planet   (PLAA-nuht)

science   (SIY-uhns)

spiral   (SPIY-ruhl)

star   (STAHR)

telescope   (TEL-uh-skohp)